W9-AMN-784

OCEAN TIDE POOL

BY
ARTHUR JOHN L'HOMMEDIEU

Children's Press
A Division of Grolier Publishing
New York London Hong Kong Sydney
Danbury, Connecticut

Childrens Press 4-7-99 $18.00
T 116412
EAU CLAIRE DISTRICT LIBRARY

Created and Developed by The Learning Source

Designed by Josh Simons

All illustrations by Arthur John L'Hommedieu

Photo Credits: Mike Bacon/Tom Stack & Associates: 9; Gerald and Buff Corsi: 22; Jeff Foot/Tom Stack: 10; J. Lotter Gurling/ Tom Stack: front cover, 6, 32; Thomas Kitchin/Tom Stack: 4-5, 17, 22 (inset), 26; Randy Morse/ Tom Stack: 1, 3, 11, 18, 19, 28, 29, back cover; Brian Parker/Tom Stack: 13 (inset), 16, 20, 27; Tammy Peluso/Tom Stack: 7, 21, 24; Milton Rand/Tom Stack: 8, 12-13; Ed Robinson/Tom Stack: 23; Therisa Stack/Tom Stack: 25.

Library of Congress Cataloging-in-Publication Data

L'Hommedieu, Arthur John.
Ocean tide pool / Arthur J. L'Hommedieu.
p. cm. -- (Habitats)
Summary: Describes how tide pools are formed and some of the animals and plants that can be found in them.
ISBN 0-516-20740-7 (lib. bdg.) 0-516-20373-8 (pbk.)
1. Tide pool ecology--Juvenile literature. [1. Tide pools. II. Series: Habitats (Children's Press)
QH541.5.S35L46 1997
577.69'9--dc21

97-17674
CIP
AC

© 1997 by Children's Press®, a division of Grolier Publishing Co., Inc.

All rights reserved. No part of this publication may be reproduced or transmitted in any form or by any means now in existence or hereafter to be devised—except for a reviewer who may quote brief passages in a review—without written permission from the publisher.

Printed in the United States of America
1 2 3 4 5 6 7 8 9 10 R 06 05 04 03 02 01 00 99 98 97

Waves flow onto the seashore—up and up, crashing and splashing again and again. They carve patterns in the sand and fill the empty places between the rocks.

Something else happens, too. As each new wave comes in, it carries along many of the plants and animals in its path.

This is high tide. For a time, water covers the whole beach, flooding everything in sight.

Soon the tide begins to ebb, returning the water to the sea. Back it goes, uncovering the sandy beach little by little. In a matter of hours, the waves are all gone and the sand is nearly dry. The only water that remains is trapped in pools between the rocks. Each of these leftover pockets of water is called a tide pool.

A tide pool is more than just water. It also is home to many animals. Some may live their whole lives here. Others may come and go with the waves. But one thing is certain. The tide pool will never again be exactly as it is right now.

The water in all the world's oceans rises and falls with the tide. The tide changes about every six hours. Twice a day the tide is high, and twice a day it is low. Between tides, the water is either slowly coming in or moving back out to sea.

Seaweed grows in the tide pool. It shades the animals from the sun and keeps them moist and comfortable during low tide.

All seaweeds are members of the algae family. They have no roots, flowers, or leaves. Instead, seaweeds grow right on the rocks using special grabbers called holdfasts. These also make a sticky glue to help them stay in place.

Like trees in a jungle, seaweed provides food and shelter for many animals. Here, a baby crab pokes out from beneath the seaweed and looks for food.

Many crabs live in the tide pool. This purple shore crab will fight other crabs for a broken clam that has washed in during high tide.

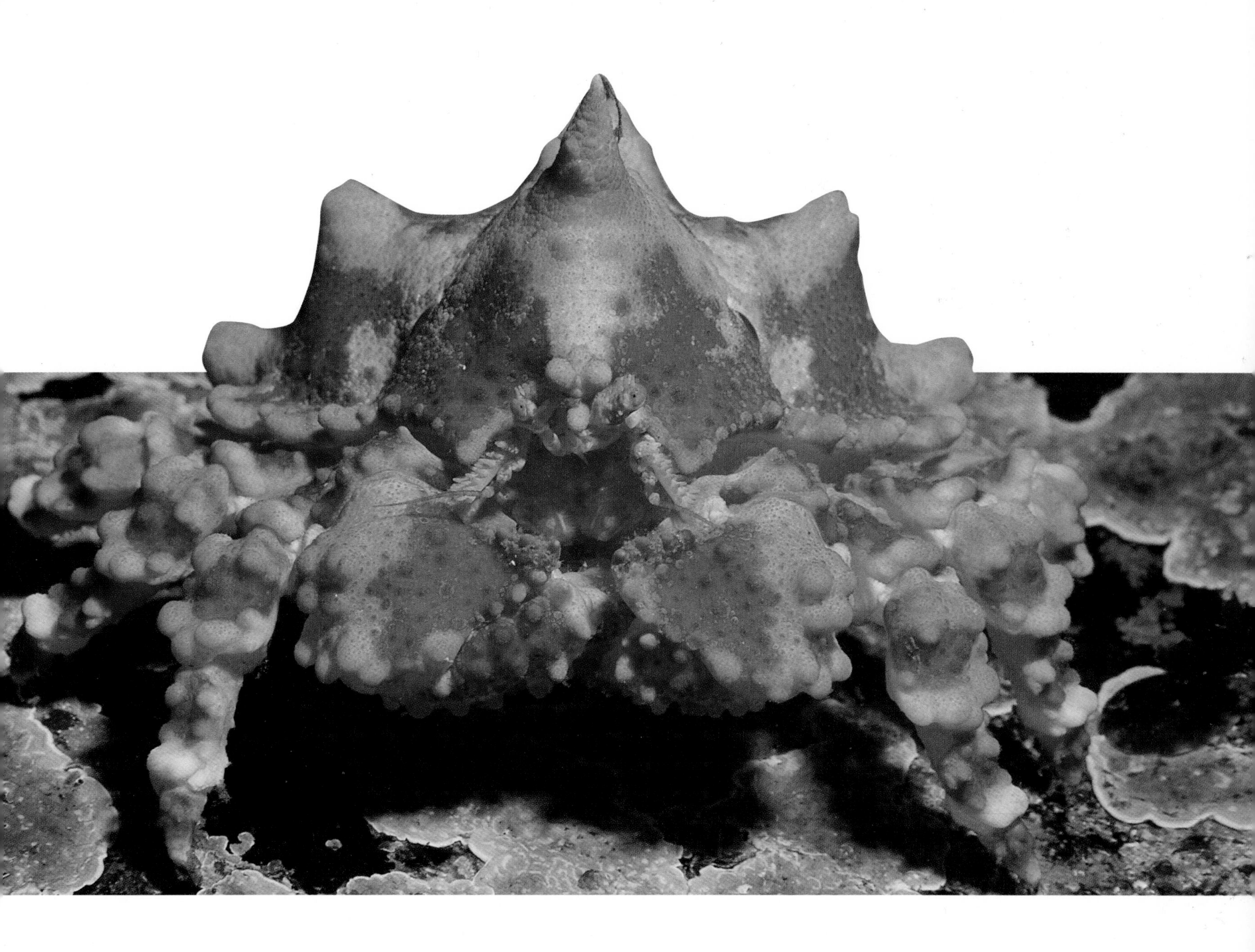

Crabs help keep tide pools clean by scavenging, or eating whatever dies there. This small king crab is surviving on tide pool food for now. But when it grows bigger it will have to move farther out to sea to find enough to eat.

Thousands of shelled barnacles live on the rocks in and around the tide pool. Although they look like little rocks themselves, barnacles are actually related to crabs. When the tide is out, barnacles close their shells to trap water inside. This helps them stay safe and damp under the burning sun.

When high tide returns, the barnacles will open their shells and get ready to feed. Their feathery legs will reach out to catch food floating in the seawater.

EAU CLAIRE DISTRICT LIBRARY

Mussels—lots of them—cover the rocks, too. Mussels are mollusks, soft-bodied animals that often live in hard shells. Some mollusks, such as mussels, clams, scallops, and oysters, live inside two shells.

Snails are mollusks, as well. But they only have one shell. Periwinkle snails use their rough tongues to feed on algae that they scrape from rocks. Larger snails, like whelks, tend to feed on smaller mollusks.

What is this moving along so quickly? It can't be a snail—it's far too fast! And it has claws, eight legs, and two eyes up on stalks. It isn't a snail at all! It's a hermit crab. Hermit crabs have no shells of their own, so they usually live inside empty snail shells. As the hermit crab grows, it seeks out newer and bigger shell homes.

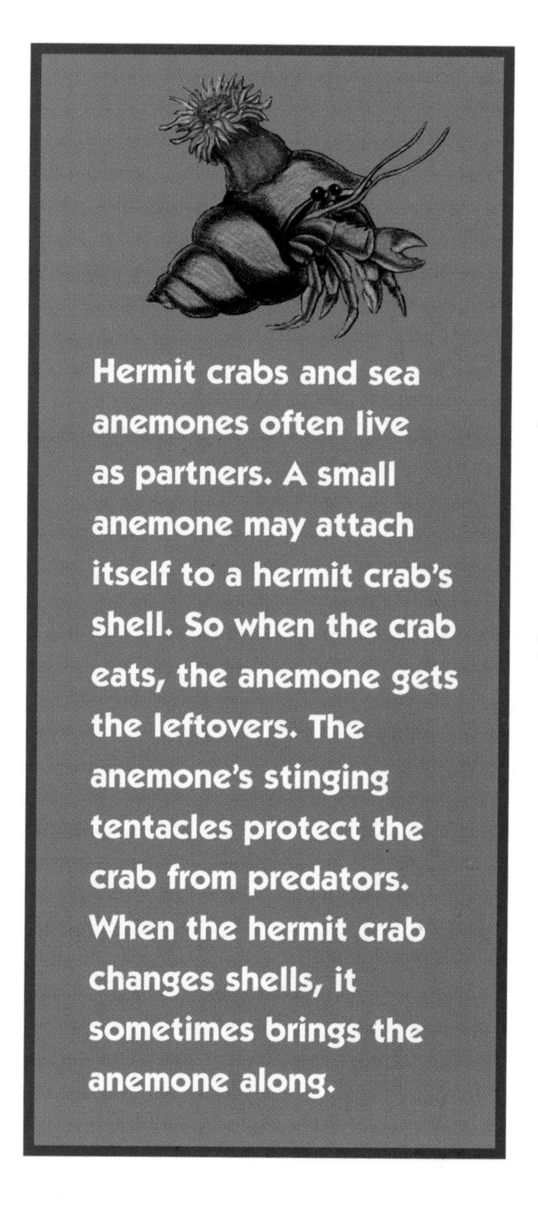

Hermit crabs and sea anemones often live as partners. A small anemone may attach itself to a hermit crab's shell. So when the crab eats, the anemone gets the leftovers. The anemone's stinging tentacles protect the crab from predators. When the hermit crab changes shells, it sometimes brings the anemone along.

At first glance these sea anemones look like lumps of jelly. But as the tide comes in, they begin to open up. Although they seem like beautiful flowers, sea anemones are actually meat-eating animals. They catch all the food they need with their tentacles.

An ill-fated starfish is trapped in the deadly tentacles of a large sea anemone. The anemone quickly stings and kills the little starfish, which provides the anemone with an excellent meal.

Unlike the baby starfish, this tiny clown shrimp is safe from the sea anemone's deadly stings. The shrimp lives right there among the tentacles. Larger animals that try to eat the shrimp often end up as dinner for the anemone.

A jellyfish is stranded in the tide pool. Although the creature looks just like a floating umbrella, it is actually made mostly of water. Jellyfish are related to sea anemones and also use stinging tentacles to catch food. Perhaps this jellyfish will be able to escape during the next high tide.

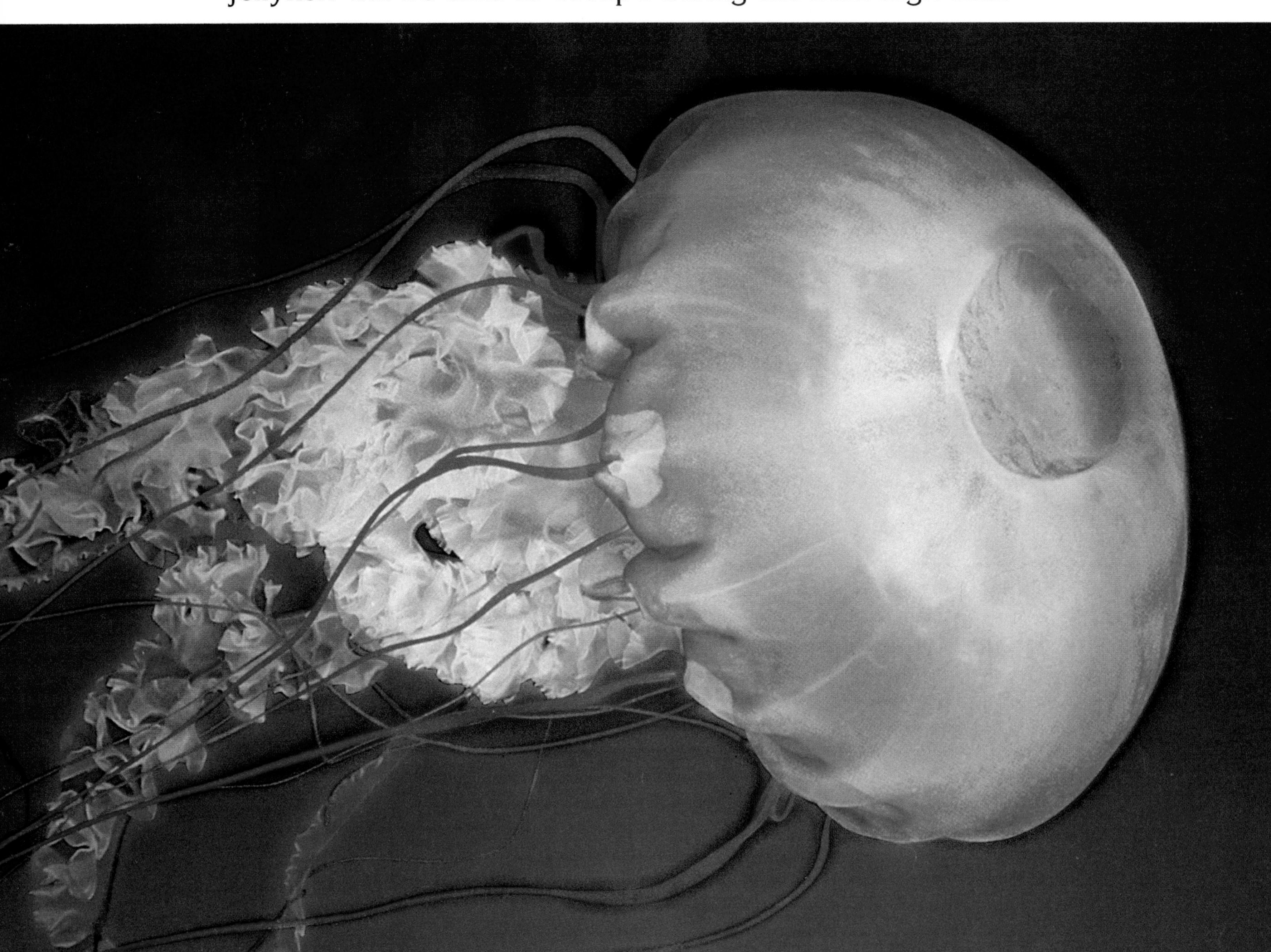

Now the tide turns, and water flows back up the beach. Starfish creep from hiding places between the rocks looking for mollusks to eat. Starfish have no heads or tails, so when they move, any arm can lead the way.

Starfish are not fish at all. They are echinoderms (ih KY nuh durmz), spiny-skinned sea creatures. Starfish have hundreds of tiny tube feet beneath their arms. Each foot has a suction cup to grip onto rocks.

This hungry starfish slowly pries open a mussel with its strong arms. Then the starfish pushes its stomach right through its own mouth and into the mussel shell. There, it surrounds the mussel and digests it, leaving the empty shell behind.

If a starfish loses an arm it will grow a new one. Even more amazing, the lost arm will grow a new starfish body. Some starfish even reproduce by pulling themselves apart and replacing their own missing pieces!

Now some prickly sea urchins come out to feed. Like starfish, they move around on special tube feet. The feet make sea urchins good climbers and help them hang on to rocks during rough weather.

Like starfish, sea urchins are also echinoderms. They search the bottom of the tide pool for tiny plants and animals. Their mouths have five toothlike plates for scraping food from the rocks.

At the bottom of the tide pool, a sea cucumber spreads its leafy tentacles. Waving in the water, these tentacles catch small creatures for the sea cucumber's dinner.

Fish—sculpins, blennies, and killifish—dart about with lightning speed. These fish are small, fast, and very hard to catch. There is plenty for them to eat in the tide pool. They find young shrimp and crabs, and some even feed on tough-shelled barnacles.

Tide pool fish are experts at camouflage. They need to be. Hungry shore birds, such as these gulls and oystercatchers, are forever hunting from above. But the rising tide has made the water deeper, and so for now the fish are safe.

The tide is high again, and with it comes a fresh supply of plankton. These tiny floating plants and animals are so small that they can only be seen with a microscope. Many tide pool animals eat plankton. Others eat the animals that eat the plankton. One way or another, plankton are food for all animals of the tide pool and the sea beyond.

While the tide is high, the tide pool is once again part of the ocean. Some creatures may swim away, and new ones might arrive. Others may be trapped here when the tide goes back out. But no matter what happens, the tide pool will always change from one tide to the next.

More About

Seaweed page 8
The round bubbles on this seaweed are called air bladders. Like tiny water wings, they help the plant float on the surface of the tide pool.

Shore Crab page 10
A crab's claws are not just for show. This one uses its big front claws for self-defense as well as for grabbing food. The claws also are useful for walking and digging.

King Crab page 11
There is good reason for this creature's name. With an adult weight of up to 12 pounds (5.4 kilograms), it truly is a king-size crab!

Barnacle page 13
Barnacles spend their lives attached to one spot. But this doesn't mean they don't get around. Besides rocks, some also grow on ship bottoms, turtle shells, and even whales.

Hermit Crab page 16
Hermit crabs go to great lengths to get new homes. They may pull a live snail out of its shell. An especially ornery crab might even yank another hermit crab out of a shell it has taken over.

Sea Anemone page 18
Some sea anemones reproduce by growing buds on their bodies, just like plants. In time the buds break off and grow on their own.

This Habitat

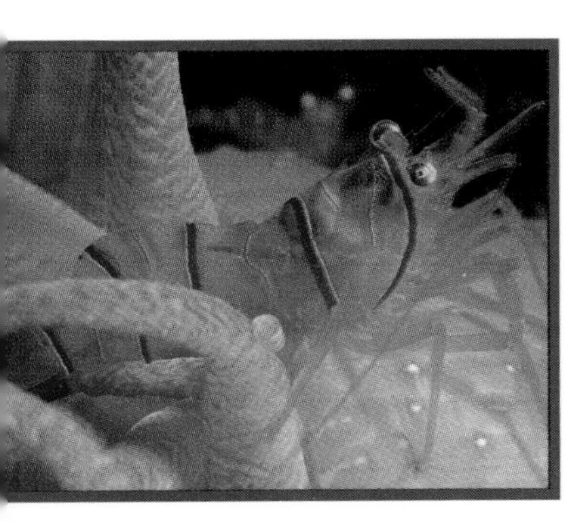

Shrimp page 19
This clown shrimp has an unusual type of camouflage. It is transparent. Both predators and prey often look right through the clown shrimp and see only what lies behind it.

Sea Urchin page 24
Even their dangerous spines are not enough protection for some sea urchins. So they camouflage themselves by covering their spines with pieces of seaweed.

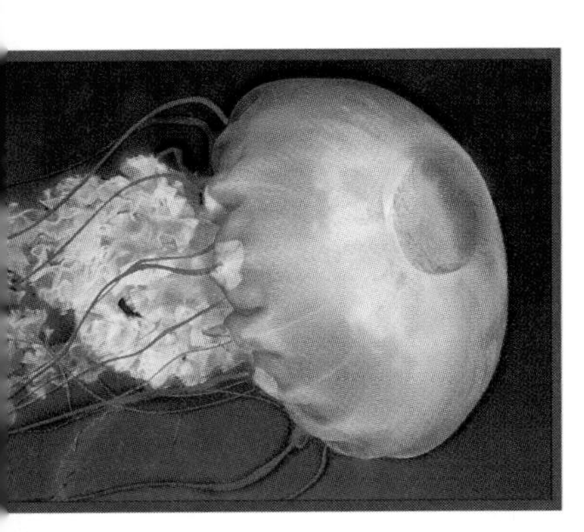

Jellyfish page 20
Jellyfish come in all sizes. Small ones are no bigger than peas. But the largest ones can grow to more than six feet (1.8 meters) across—bigger than many tide pools!

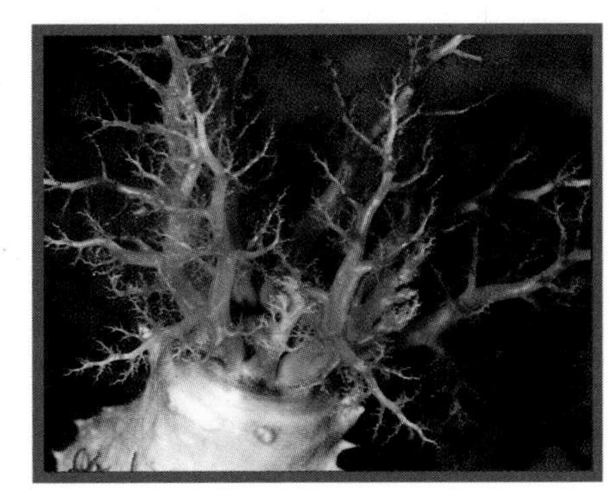

Sea Cucumber page 25
When threatened, some sea cucumbers send sticky, threadlike tubes out from their bodies. These tubes keep attackers from getting too close.

Starfish page 21
Starfish "see" through small colored eyespots at the tips of their arms. Each eyespot senses light but does not actually see images.

Oystercatcher page 26
An oystercatcher's bill is sharp and shaped like a chisel. This makes it very useful for opening the shells of oysters, clams, and other mollusks.

EAU CLAIRE DISTRICT LIBRARY